AF745351

WITHDRAWN
University of
Reading Library

90 0385156 3
TELEPEN

Ministry of Agriculture, Fisheries and Food

Possible patterns of agricultural production in the United Kingdom by 1983

London: Her Majesty's Stationery Office

© Crown copyright 1979
First published 1979

385,156

CAL

ISBN 0 11 240326 3

Possible patterns of agricultural production in the United Kingdom by 1983

Introduction

1. This note presents the results of a study of two possible patterns of agricultural production in the United Kingdom some five years hence. The study was undertaken in connection with the preparation of the White Paper "Farming and the Nation". The analysis is based on projections of yields, assumptions concerning improvements in the efficiency of resource use and alternative levels of producers' real prices. Likely developments in the demand for food are also taken into account. The projected patterns of production are compared with the position in recent years and details of the assumptions required for an exercise of this kind are given in order to facilitate further discussion of the future development of the agricultural industry.

2. Only two from among a wide range of possible future production patterns are considered. The scope for variation is considerable, since agricultural production covers some ten main commodity groupings (and each of these may include numerous and often quite different commodities). Also, the amount of land available for farming is declining, and expansion of the area for one commodity pre-empts the land and other resources that might be used for another. Since changes tend to take place relatively slowly, it is likely that the pattern of output in 1983 will be influenced by the existing pattern. It will also depend on the scope for improvements of a technical kind (eg in yields); on the development of systems leading to improvements in input/output ratios; and on relative commodity real price and profitability levels.

3. The assessments which are presented are not intended to be forecasts of what will happen nor should they be interpreted as norms that should be achieved regardless of the economic circumstances. The aim is much more limited. It is to provide broad indications of what, seen from the present production base, may be reasonably expected to happen provided certain assumed technical and real price changes materialise, and assuming that no serious disruptions arise during the projection period. The projected figures should not be taken to be definitive and are put forward to provide a basis for further consideration of production prospects.

Assumptions

4. Any projection of future agricultural production must involve assumptions about the rate of technical change (such as changes in the yields of crops and livestock

products, and of feed to output conversion rates), and about economic factors such as producer prices and cost changes. For this study the starting point was to project a pattern of United Kingdom agricultural production, taking account of the technical possibilities and, for example, changes in livestock numbers already in train. This yielded the production projection based on the assumption of constant real prices. The projections obtained in this way were then adjusted to take account of the alternative assumption of declining real prices. The two possible production patterns derived in this way are presented below. These projections are not meant to simulate what is likely to take place over the next five years but it is hoped that they will provide a basis for comparison with alternative assessments made, possibly by organisations concerned with the future development of the industry.

5. For the first set of projections it is assumed that, in general, producers' real prices remain unchanged to 1983 at their 1977 levels. In other words, prices received by producers are taken to increase in money terms at broadly the same rate as increases in the general level of prices in the economy. For the second and alternative set, it is assumed that producers' real prices, in general, decline by 1% per annum from 1977 to 1983, ie by just over 6% over the period. In the past, real prices have fluctuated and the movements have varied, often to a marked extent, between the various commodities; but since the mid-1950s there has been a decline of about 1% per annum on average. It needs to be re-emphasised that the general price assumptions are not intended to convey judgments about how agricultural real prices are likely to change in practice over the next six years. Exceptionally, in the case of poultrymeat, eggs and potatoes, the method of projection, which is outlined below, precludes the need to make specific assumptions about producers' real prices.

6. The assumptions relating to the possible technical developments over the period to 1983 were made after considering progress over the last decade, incorporating advice from technical experts within the Ministry of Agriculture, Fisheries and Food and taking account of the views of outside organisations. Essentially the projections assume the continuing steady introduction of new products and techniques and the wider application of recent developments. Improvements in livestock stocking rates, as well as increases in the yields of crops and livestock, have need assumed.

7. For this exercise, agricultural commodities are divided into two general categories. First, there are activities such as most field and horticultural crops, sheep, beef and dairying, for which the projections are made by following through the implications of the technical and price assumptions, but in the case of livestock, taking account of production changes over the next two years which are considered to be already in train. For poultrymeat, eggs and potatoes, however, the projections are based on "market clearing" models. Models of this type assume that whatever quantity of product comes on the market will be absorbed, if necessary, by means of a price adjustment. Such models are appropriate for commodities for which the production cycle is relatively short, for which we are virtually self-sufficient and for which prices can fluctuate significantly. For pigmeat the method needs to allow for the more pronounced cyclical nature of production, and for the competing requirements of the pork and bacon markets. The projections for individual commodities take account of the overall constraint on production imposed by the availability of land for agriculture and, where relevant, inter-relationships between commodities.

8. Although the projected 1983 production indices shown for the various commodities are based on 1977 final figures, the 1978 provisional estimates are included in all tables for comparative purposes and have been taken into consideration in making the commodity projections themselves.

The projections

Total Agricultural Area

9. The rate of loss of agricultural land depends upon the demand for land for non-agricultural purposes and policies which affect the transfer of land from farming to other uses. As a result of urban and industrial development, and the creation of new recreational areas, roads and forests, the area of agricultural land is declining. Between 1971 and 1978 the total area included in agricultural holdings—not, therefore, including common rough grazings—fell, on average, by about 50,000 hectares per year, and this has limited both the scope for expansion of the arable area and the stock-carrying capacity of the farming sector. Any projections of the scale of agricultural output need to rest on assumptions about the available land area, and in this study the continuation of an average loss of 50,000 hectares per annum has been assumed, implying a total loss of 250,000 hectares over the five year period from 1978 to 1983. This assumption has implications for the total area devoted to tillage including, in particular, the cereals and grassland areas. The projected grassland and rough grazing areas also affect the implied stocking rates. A summary of crop areas is given in Table 1.

Milk and Beef

10. The projections relating to the *dairy sector* are considered first since the size of the dairy herd influences the output of beef through the projected supplies of dairy calves and culled dairy cows. In assessing prospects for the size of the dairy sector account is taken of assumed returns from the sale of calves and culled cows. With constant real prices, milk yield per cow is assumed to increase from the 1977 level of 4,452 litres/cow to about 5,060, ie an increase of about 2% per annum. It should be noted, however, that the average yield in 1978 is estimated to be high relative to 1977 so that the trend increase assumed for the years 1978 to 1983 is about 1¾% per annum. A summary of milk and other yields is given in Table 2 and is also shown on graphs 1–4. In projecting the size of the United Kingdom dairy cow herd in 1983, the latest estimates of dairy cow numbers have been used and these are assumed to reflect the full effects of price changes to date. Taking these factors into account, on the constant real price assumption the dairy cow numbers at June 1983 are projected at 3,335 thousand, about 2% higher than in June 1977 and than the provisional figure for June 1978. A summary of livestock numbers is given in Table 3. With yield per cow projected to be nearly 14% higher than in 1977 the overall volume of milk production in 1983 compared with 1977, is projected to be higher by about 2,280 million litres, or almost 16%. On the assumption that real prices decline by 1% per annum, dairy cow numbers at June 1983 are projected to be 3,200 thousand, while average yields per cow are projected at 4,970 litres, somewhat lower than the projected yield at constant prices, but still over 11% higher than in 1977 and 7½% more than the estimated yield for 1978. The overall volume of milk production, compared with 1977,

is projected to be higher by 1,325 million litres or 9%. A summary of production levels is given in Table 4.

11. Constant real prices would involve little change in the total consumption of milk in liquid and manufactured form. This is because the slight downward trend in demand is largely offset by a small positive income elasticity. Consumption of cheese has been rising in recent years, due to an upward trend reinforced by some income effect, and this is expected to continue. Under the assumption of constant real prices, consumption of butter, on the other hand, is expected to fall slightly as income effects are more than offset by a downward trend in demand. Consumption of milk in the form of other manufactured products is not expected to change significantly over the period. With declining real prices, although butter consumption might show some increase, the total consumption of milk in all forms is not expected to be affected appreciably because the price elasticity of demand for most milk products is relatively low.

12. The size of the United Kingdom *beef* breeding herd has been declining since 1975. If producer prices remain constant in real terms, it is considered that this decline would slow down significantly and stability soon be achieved. In making the projections a price elasticity of supply (in terms of beef cow numbers) of 1·2 was used, but it was also assumed that the response to a price change would take three years to complete. In other words, the response to, say, a 1% real price fall would be a 1·2% reduction in the number of beef cows, seen as a 0·4% reduction in each of three years. The projections have also taken into account latest short-term forecasts of the size of the beef cow herd to 1980, and these are considered to reflect adjustments to changes in producer returns up to the end of 1977. On this basis, numbers of beef cows in 1983 are projected at 1,570 thousand, just under 1% below the provisional figure for 1978. With declining real prices the number of beef cows is projected at 1,510 thousand, about 5% below the provisional 1978 level.

13. The volume of beef production is dependent on the number of calves fattened from the beef and the dairy herds, average carcase weights, the number of cull cows slaughtered, and the volume of beef produced from imported store cattle. The number of calves to be reared for beef will depend on the level of total calf disappearance (ie calf slaughterings plus exports) and those reared for breeding replacements. From the smaller beef herd projected for 1981 with constant real prices, some 1,070 thousand calves are projected to be available for fattening; these, together with calves from the dairy herd but excluding calves slaughtered and exported, give a projected total of 2,720 thousand fat cattle available for slaughter in 1983. On the assumption of declining real prices, the projected number would be smaller at about 2,650 thousand. Average carcase weights have been projected to increase by about 1 kg per head per annum after taking into account the latest estimates for 1978. Based on these various assumptions and allowing for the projected volume of beef from cull cows and imported store cattle, production in 1983, given constant real prices, is projected at 1,070 thousand tonnes. This is some 4% above the level in 1977. It is slightly lower than the Annual Review estimate of 1978 production, but there are indications that the final out-turn for 1978 production may be somewhat lower than the forecast given in the Annual Review White Paper. With declining real prices production is projected at

about 1,045 thousand tonnes, showing a 1% gain from the 1977 level and a small reduction on the 1978 provisional estimate.

14. Beef consumption at constant real prices is projected to increase by about 5%, due almost entirely to the effects of assumed higher real incomes; with declining real prices, consumption is projected to increase in total by some 7%. These projections follow from the assumptions about real price movements. In the event, the interaction between the supply and demand for beef and other meats would determine real price levels which may differ from those assumed. Therefore firm conclusions cannot be drawn about the future domestic share of the United Kingdom beef market from the above output projections.

Mutton and Lamb

15. The projections of mutton and lamb production assume, as in the case of beef, that the latest available forecasts of the size of the breeding flock in 1979 reflect the full response to producer price changes that have taken place up to the present, and that there will be no significant change over the period to 1983 in lambing rates and in average carcase weights. With producer prices remaining constant in real terms the total breeding flock is projected to rise to 14,485 thousand head by 1983 which, compared with 13,702 thousand in 1977 and 14,149 thousand in 1978, represents increases of about 6% and 2% respectively. Mutton and lamb production in 1977 was somewhat below the level that might have been expected given the size of the breeding flock. This reflected retention of breeding stock and late marketings, and a carryover of these into 1978. The projection for 1983 assumes the more normal relationship between the breeding flock and production and accordingly the latter is projected to grow more rapidly than the breeding flock during the five-year period 1977–1983. The increase is put at about 12%, implying production of some 255 thousand tonnes, compared with 229 thousand tonnes in 1977. On the other hand, with declining prices the lowland breeding flock is assumed to fall by some 1·35% for every 1% fall in producers' real price, the adjustment being spread over three years, while the upland breeding flock is assumed to remain static once responses to present price changes have been exhausted. With declining prices the overall size of the United Kingdom breeding flock is projected to be 14,130 thousand head by 1983. On this basis, mutton and lamb production would be about 250 thousand tonnes in 1983—an increase of some 9% compared with 1977.

16. The trend in demand for lamb has been steadily downwards for a number of years. This trend is expected to continue and the effect on consumption is only partly offset by the effect of higher incomes. Consumption at constant real prices is projected to fall by almost 5% by 1983, and by 3% assuming declining real prices.

Cereals

17. The area planted to cereals is projected to increase from 3,706 thousand hectares in 1977 to 3,850 thousand hectares in 1983 at constant real prices, and to 3,825 thousand hectares with declining real prices. The projected area for 1983 is similar to the provisional 1978 area and represents an increase of about 4% on 1977 at constant real prices and 3% with declining real prices. Cereal yields have been on a rising trend

and this is projected to continue over the coming years. At constant prices, the likelihood of further improvement in wheat yields is greater than for other grains because of intensive research in this field. Thus a growth in wheat yield of 0·08 tonnes per hectare per annum over 1977 is assumed—appreciably higher than the growth in yields projected for other cereals.

18. The area sown with wheat tends to fluctuate from one year to the next, as does the average yield, depending on cultivation conditions in the autumn when most wheat is planted. In the autumn of 1976, sowing conditions were not good and consequently the area of wheat in 1977 was below trend. It recovered in 1978 following the excellent autumn cultivation weather in 1977. Assuming normal weather conditions and constant real prices, the area planted with wheat is projected to rise by 1983. The combined area of barley, oats and other cereals is projected to be just under its 1977 level. Applying the assumed yields to these projected areas implies, at constant real prices, a production of some 18,660 thousand tonnes of cereals by 1983. This is about 11% higher than the actual crop harvested in 1977, but only some 7% higher than the forecast production for 1978. Slightly lower yields and area are projected with declining real prices, implying a production of 18,265 thousand tonnes in 1983. This would be about 9% higher than production in 1977 but only some 5% higher than estimated production for 1978.

19. Total demand for cereals includes requirements for direct human consumption and for animal feed. Since consumer demand for cereal products is not very sensitive to small price changes, a broadly similar change is expected under both real price assumptions. The demand for wheat used for milling is projected to fall by about 3%, but the demand for barley which is required for brewing and distilling is forecast to increase by 8% by 1983. The livestock projections suggest only a small change in the demand for cereals for animal feed.

Sugar Beet

20. The sugar beet area has been showing an upward trend and it is assumed, with constant prices, that it will rise further to about 230 thousand hectares in 1983, compared with 202 thousand hectares in 1977 and 210 thousand hectares in 1978. The yield of refined sugar per hectare will depend on the yield of beet, its sugar content and the efficiency of the extracting processes. The yield of refined sugar per hectare from the 1977 crop was 4·72 tonnes per hectare and is estimated at 4·9 tonnes per hectare in 1978. Production of refined sugar amounted to almost 950 thousand tonnes and around 1,030 thousand tonnes in 1977 and 1978 respectively. Yields of refined sugar from sugar beet have been depressed in recent years, but improvements are expected. It has been assumed that yields will be around 5·2 tonnes of refined sugar per hectare by 1983. On this basis, which allows for the improvement in extraction rates expected as a result of the British Sugar Corporation's modernisation plan, refined sugar production would amount to about 1·2 million tonnes in 1983, an increase of about 26% over the quantity obtained in 1977 and 17% more than the provisional 1978 level. With declining real prices the projected sugar beet area is 215 thousand hectares by 1983 and refined sugar production would be just over 1·1million tonnes, or about 18% and 10% more than in 1977 and 1978 respectively.

21. In terms of refined sugar, demand at constant real prices is expected to fall by about 4% by 1983. This overall decline reflects two opposing influences, namely, demand for packeted sugar which is projected to fall by 11%, and manufacturing demand for sugar which is expected to increase somewhat. Consumption with declining real prices is also expected to fall from current levels: the underlying trend would more than offset the effects of the lower price.

Horticulture

22. Since this sector comprises numerous fruit, vegetable and non-food crops, projections of the area devoted to it, and of output, are presented in aggregate terms. With constant real prices, total horticultural area is projected at 295 thousand hectares in 1983. The long-term trend is for the volume of output to rise and it has been assumed that this growth will continue at about 1% per annum over the next few years. However, since the volume of output was well below trend in 1977, an increase of 12% in output, at constant 1977 prices, is implied between 1977 and 1983.

23. It is difficult to assess the likely changes in horticultural production if declining real prices are assumed. The sector is a diverse one and responsiveness to changes in real returns will vary considerably by type of crop. These range from orchard fruits and glasshouse crops, (for which considerable investment has already been committed and where supply adjustments may be expected to be spread over a long time-span), to field vegetables where adjustments can be made rapidly. Overall, a decline in real prices of 1% per annum to 1983 would be expected to reduce somewhat the projected cultivated area, particularly that devoted to the production of field vegetables. The output projections with declining real prices show an increase of 8% over 1977.

24. Demand for fruit and vegetables is expected to rise by about 2% by 1983 given constant real prices, and by a little more than this if real prices decline. Within this overall change the effect on expenditure is likely to vary considerably by type of product.

Pigmeat

25. Pigmeat production in the United Kingdom, as in most countries, has been characterised by significant cyclical movements. Production peaks can be identified in 1971/72, 1973/74 and 1977, with troughs in the intervening years and in 1978. In these circumstances any projection of pigmeat production cannot purport to relate to a particular year, especially in view of the variation which exists between cycles both in respect of their amplitudes and lengths. The output projection, therefore, must be taken as providing only an indication of the possible average production level in the early 1980s.

26. Of the total pigmeat consumed in the United Kingdom, roughly 40% is in the form of bacon, and 30% each as fresh pork and other pigmeat products such as sausages, pies and pork luncheon meat. In recent years some 30% of domestic pigmeat production has been used for bacon and ham; of the remaining 70%, about half is consumed as fresh pork and the remainder as pork products. In the United Kingdom in the early 1980s the demand for fresh pork, at constant real prices, is expected to increase by about 10%, but that for bacon and most other pigmeat products (including

imported produce) is expected to decline. For bacon the decline is put at 2½% and for the other products at about 5% by 1983. Thus the demand for all pigmeat, given constant prices, is expected to show little change over the period. A decline in real prices would not affect consumption significantly as the real prices of other meats would also be falling.

27. Given these projections of total market requirements in the United Kingdom for pigmeat in the early 1980s and allowing for imports, the analysis suggests, with constant real prices (implying also constant real cereal feed prices—the main cost item in pigmeat production), a domestic pig breeding herd of about 895 thousand, considerably above that in 1977 and 1978. Total pigmeat production from a herd this size would be about 940 thousand tonnes, an increase of nearly 4% and 8% over 1977 and the 1978 (forecast).

28. Declining real prices imply a reduction in the price of cereal feed (and thus pigmeat production costs), lower producer returns and lower prices of competing meats. The reconciliation of output and demand suggests a United Kingdom pig breeding herd in the early 1980s of about 865 thousand, somewhat below the 1976 level but 4% and 3% above the 1977 and provisional 1978 estimates. Production of pigmeat from a herd of this size would be about 910 thousand tonnes, about the same as in 1977 and 4% more than in 1978.

Poultrymeat, Eggs and Potatoes

29. The United Kingdom is virtually self-sufficient in these products and their production cycles are relatively short. Because of this, it seemed, as has already been mentioned, more appropriate to use "market clearing" models. These attempt to simulate the supply and demand adjusting processes of the market-place, rather than to project solely on the basis of likely technical developments and on assumptions about the producer price. The "market clearing" approach assumes that changes in production are kept in line with changes in demand by means of price adjustments. The assumed rates of adjustment of both production and consumption to real price changes are the key factors in models of this sort, but the projections are also dependent on assumptions of productive efficiency and any underlying trend in demand.

30. *Poultrymeat* production rose rapidly up to the early 1970s, but the rate of increase has slowed down in recent years as is indicated in Table 4. Consumption, which is largely governed by a strong upward trend, is projected to increase further to 1983. The volume of production required to meet this demand, assuming constant real prices (including the price of cereal feed, the most important cost item), is put at 750 thousand tonnes, an increase of 11% and 8% over the 1977 and 1978 forecast production figures respectively. In the case of *eggs*, production since 1970 has shown some fluctuation around a slightly downward trend. The demand for eggs is projected to decline marginally and production in 1983, at constant real prices, is projected to be about 2% to 3% below the levels in 1977 and 1978.

13. With declining real prices, the cost of cereals, and hence the implied costs of production, would decline for both poultrymeat and eggs. However, the price of competing foods, such as other meats and proteins, would also be lower. In effect, therefore,

the levels of consumption and production of these commodities by 1983 are projected to be essentially the same under both real price assumptions.

32. Given the uncertainty at present surrounding the precise form of an EEC regime for *potatoes*, the 1983 output projections for this crop have also been made on the assumption that domestic supply is adjusted to meet changes in demand and that the United Kingdom remains virtually self-sufficient in respect of main-crop potatoes. Further concentration of production amongst growers, together with technical advances, are expected to lead to increases in yields from about 28½ tonnes per hectare in 1977 to over 34½ tonnes per hectare in 1983. Given projected yields and assuming a continued low price elasticity of demand for potatoes (including potato products), production and consumption would be about 6,680 thousand tonnes in 1983, an increase of about 1% on the 1977 level but a reduction of about 2% from the 1978 forecast. The area of potatoes implied would be 194 thousand hectares. The projected levels of output and overall consumption remain virtually the same under both price assumptions; an increase in the consumption of processed potato products is expected to offset a decline in that of fresh potatoes.

Overall assessment of the projections

33. It has been indicated above that the commodity projections were made taking account of the relationships between products and the limitations on growth imposed by, for example, the assumption on the declining land area available for farming purposes. The implications for the area devoted to individual crops, including grassland, are shown in Table 1. The decline in grazing area is not matched by a proportionate decline in the numbers of grazing livestock, so that higher stocking densities would clearly have to be achieved for the projections to have validity. However, under the assumption of declining real prices there would be virtually no change in stocking rates between 1978 and 1983.

34. The aggregate gross output, gross input and net product indices (1975=100) for agriculture are shown in Table 5 for the period 1967 to 1978 (forecast) and projected to 1983. To arrive at these annual indices the figures for individual commodities and inputs were valued at 1975 constant prices. The agricultural industry was seriously disrupted by the droughts of 1975 and 1976 and this is reflected in the large declines in the volume of agricultural net product in those years (see graph 5). Some of the consequences of these setbacks were carried over into 1977, when as Table 5 suggests, in spite of a sharp recovery, net product was still somewhat below the trend level. Production in 1978 no longer reflects the consequences of the setbacks of earlier years, but as the figures for 1978 are still provisional, the rates of growth quoted in the following paragraphs are expressed as percentage increases over the average for 1977 and 1978 rather than on the 1978 figure only.

35. The overall result, assuming constant agricultural prices in real terms, is that the volume of agricultural gross output is projected to increase by about 9% above the 1977/78 level. The improvements in the efficiency of agricultural production assumed are, in general, reflected in the lower usage of inputs per unit of output. Thus, with constant real prices, the overall volume of gross input into agriculture

by 1983 is projected to be only some 4% over the 1977/78 level. These projections, all valued at constant prices imply a volume increase of about 18% on the 1977/78 level of agricultural net product (which is the difference between gross output, and gross input and depreciation).

36. With declining real prices, the projected growth of gross output of 5% over the period is associated with an increase of almost 3½% in the volume of inputs used. Thus the increase in net product implied is 10½% over the period 1977/78 to 1983.

37. The foregoing projections provide two views on the possible pattern of agricultural production in the early 1980s based on certain specified assumptions about the rate of technical change and the development of output and input prices. In the event, agricultural production each year will be affected by weather and other factors and thus may in turn affect the eventual pattern of agricultural production some five years hence. Also the growth of agricultural net product is unlikely to be smooth. The projections do however, provide points of reference for alternative and subsequent studies into the development of the agricultural industry which may be undertaken by organisations interested in the future progress of the farming sector.

TABLE 1 Crop areas at June of each year

'000 hectares

	1974	1975	1976	1977	1978 provisional	1983 (projected) At constant real prices	1983 (projected) With declining real prices
Wheat	1,233	1,034	1,231	1,076	1,263	1,275	1,275
Barley	2,214	2,345	2,182	2,400	2,360	2,375	2,350
Oats	253	232	235	195	181	175	175
Other cereals	48	42	37	35	27	25	25
Total cereals	3,747	3,653	3,685	3,706	3,830	3,850	3,825
Potatoes	215	204	222	232	212	194	194
Sugar beet	195	198	206	202	210	230	215
Oilseed rape	25	39	48	55	65	80	78
Horticulture (total)	285	285	289	302	290	295	285
Other crops and bare fallow	371	437	371	366	335	340	335
Total tillage	4,838	4,816	4,821	4,863	4,942	4,989	4,932
Temporary grass and lucerne	2,316	2,138	2,154	2,124	2,102	2,231	2,167
Total grass	7,236	7,212	7,235	7,127	7,077	6,859	6,865
Total grass and tillage	12,074	12,028	12,056	11,990	12,019	11,848	11,797
Rough grazing(*c*)	6,564	6,555	6,513	6,400(*b*)	6,330(*b*)	6,212(*b*)	6,263(*b*)
Other(*a*)	372	395	419	451	509	550	550
Total area	19,010	18,978	18,987	18,840(*b*)	18,860(*b*)	18,610(*b*)	18,610(*b*)

Notes: (*a*) Buildings, roadways, small woods.
(*b*) Includes sheep stock clubs in Scotland.
(*c*) Includes common rough grazing.

Source: June Agricultural Returns for the United Kingdom.

TABLE 2 Estimated average yields of crops and livestock products 1974–1978 and projections for 1983

	1974	1975	1976	1977	1978 forecast	1983 (projected) Constant real prices	1983 (projected) Declining real prices
Crops							
Wheat (tonnes/hectare)	4·97	4·34	3·85	4·90	5·22	5·40	5·28
Barley (tonnes/hectare)	4·12	3·63	3·51	4·39	4·24	4·6	4·55
Oats (tonnes/hectare)	3·77	3·42	3·25	4·06	4·02	4·24	4·20
Potatoes (tonnes/hectare)	31·6	22·3	21·6	28·5	31·9	34·5	34·5
Sugar (tonnes/hectare)(*a*)	2·93	3·25	3·39	4·72	4·9	5·24	5·24
Oilseed rape (tonnes/hectare)	2·1	1·7	2·3	2·6	2·5	2·6	2·6
Livestock products							
Milk (litres/cow per year)	3,925	4,102	4,267	4,452	4,621	5,060	4,970
Eggs (No./bird)(*b*)	229·5	229·0	238·5	240·5	241·5	254	254

Notes: (*a*) Refined sugar per crop hectare.
(*b*) Eggs per laying bird, including breeding flock.

Source: Annual Review of Agriculture for 1979.

TABLE 3 Livestock numbers in the United Kingdom 1974–1978 and projections for 1983

'000s

	1974	1975	1976	1977	1978 Provisional	1983 (projected) At constant real prices	1983 (projected) With declining real prices
Total cattle and calves	15,203	14,717	14,069	13,854	13,660	13,710	13,190
of which: Dairy cows	3,394	3,242	3,228	3,265	3,278	3,335	3,200
Beef cows	1,887	1,899	1,764	1,680	1,584	1,570	1,510
Total sheep and lambs	28,498	28,270	28,265	28,104	29,618	29,700	28,975
of which: Ewes	11,192	11,279	11,298	11,215	11,415	14,485 (Ewes and Shearlings)	14,130 (Ewes and Shearlings)
Shearlings	2,673	2,471	2,369	2,487	2,734		
Total pigs	8,544	7,532	7,947	7,736	7,764	8,330	8,060
of which: Sows in pig and other sows for breeding	783	710	747	725	721	765	740
Gilts in pig	107	104	137	103	122	130	125
Total poultry	139,672	136,572	142,222	134,286	134,177	134,723	134,723
of which: Table fowls (including broilers)	56,701	56,708	61,325	56,153	53,728	59,830	59,830
Laying fowls	49,924	49,359	49,085	49,119	50,744	46,835	46,835
Growing pullets	18,958	18,195	18,383	16,341	16,678	15,345	15,345

Source: June Agricultural Returns for the United Kingdom.

TABLE 4 Agricultural production in the United Kingdom 1974–1978 and projections for 1983

'000 tonnes (unless otherwise stated)

	1974	1975	1976	1977	1978 forecast	1983 (projected)(*a*)			
						Production		Approximate index (1977 = 100)	
						At constant real prices	With declining real prices	At constant real prices	With declining real prices
Milk (m. litres)	13,319	13,343	13,819	14,595	15,191	16,875	15,920	116	109
Beef	1,086	1,219	1,069	1,032	1,079	1,070	1,045	104	101
Mutton and lamb	253	264	248	229	248	255	250	112	109
Pigmeat	987	824	850	912	876	940	910	104	100
Poultrymeat	652	628	663	678	692	750	750	111	111
Eggs (m. dozen)	1,148	1,116	1,149	1,156	1,168	1,130	1,130	98	98
All cereals	16,382	13,936	13,263	16,727	17,425	18,660	18,265	112	109
Sugar(*b*)	568	641	695	949	1,025	1,200	1,125	126	118
Potatoes	6,791	4,551	4,789	6,621	6,794	6,680	6,680	101	101
Oilseed rape	53	67	111	142	162	210	200	148	123
Horticulture (£m.)(*c*)	468	528	593	712	709	795	770	112	108

Notes: (*a*) Production figures rounded to nearest 5, indices to nearest whole point.
(*b*) Refined basis.
(*c*) Valued at current prices except 1983, which is valued at 1977 prices.

Source: Annual Review of Agriculture 1979.

TABLE 5 Indices of gross output, gross inputs and net product, at constant prices

(1975 = 100)

	Gross Output(*a*)	Gross Input(*b*)	Net Product(*c*)
1967	97	94	103
1968	97	96	100
1969	98	98	102
1970	100	101	101
1971	105	102	112
1972	105	104	111
1973	106	103	113
1974	104	99	114
1975	100	100	100
1976	97	101	91
1977	107	103	115
1978 forecast	110	102	122
1983 projection at constant real prices ...	118	107	140
1983 projection with a 1% fall in real prices	114	106	131

Notes: (*a*) The Index of Gross Output indicates changes in the volume of agricultural produce, stock and work-in-progress.

(*b*) The Index of Gross Input indicates changes in the volume of material inputs purchased, but excludes depreciation, labour, rent and interest.

(*c*) The Net Product is the difference between Gross Output, and Gross Input and depreciation.

ESTIMATED AND PROJECTED ANNUAL AVERAGE YIELDS OF MILK IN THE UNITED KINGDOM **GRAPH 1**

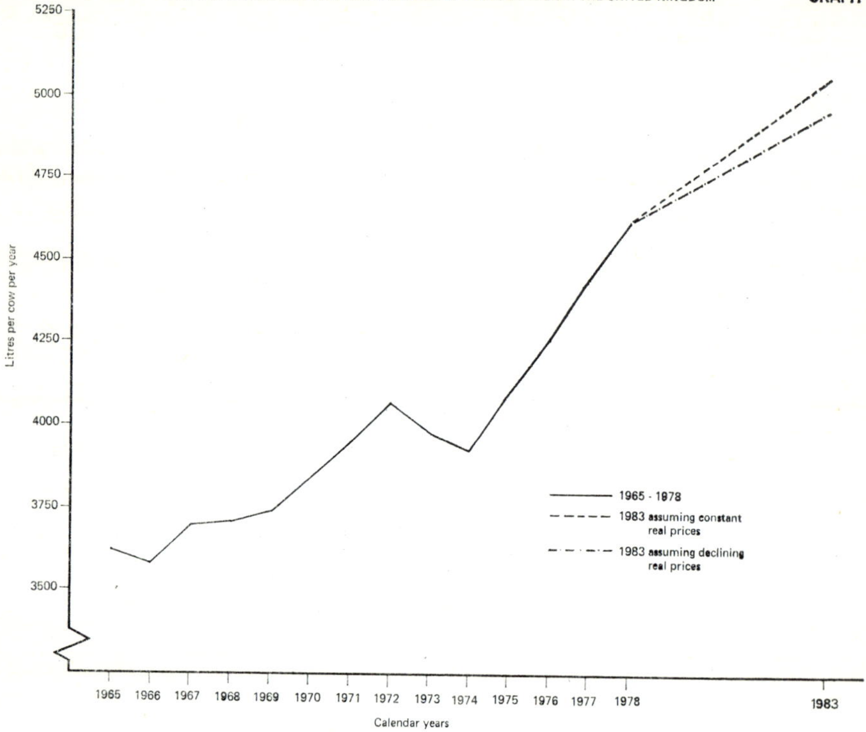

GRAPH 2

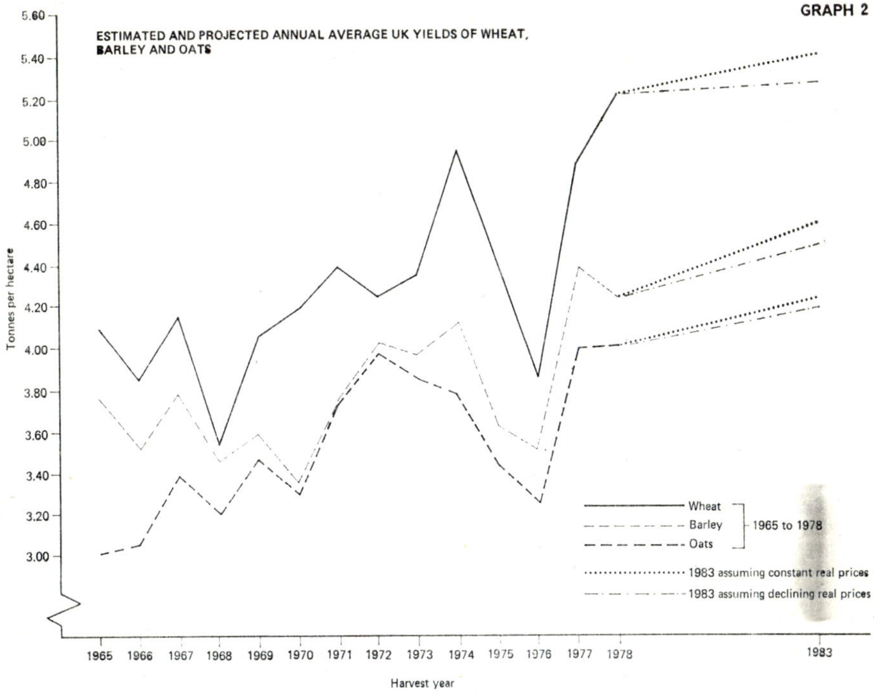

ESTIMATED AND PROJECTED ANNUAL AVERAGE UK YIELDS OF WHEAT, BARLEY AND OATS
5.60
5.40
5.20
5.00
4.80
4.60
4.40
4.20
4.00
3.80
3.60
3.40
3.20
3.00
Tonnes per hectare
Wheat
Barley
Oats
1965 to 1978
1983 assuming constant real prices
1983 assuming declining real prices
1965 1966 1967 1968 1969 1970 1971 1972 1973 1974 1975 1976 1977 1978 1983
Harvest year

GRAPH 3

ESTIMATED AND PROJECTED ANNUAL AVERAGE UK YIELDS OF REFINED SUGAR PER CROP HECTARE

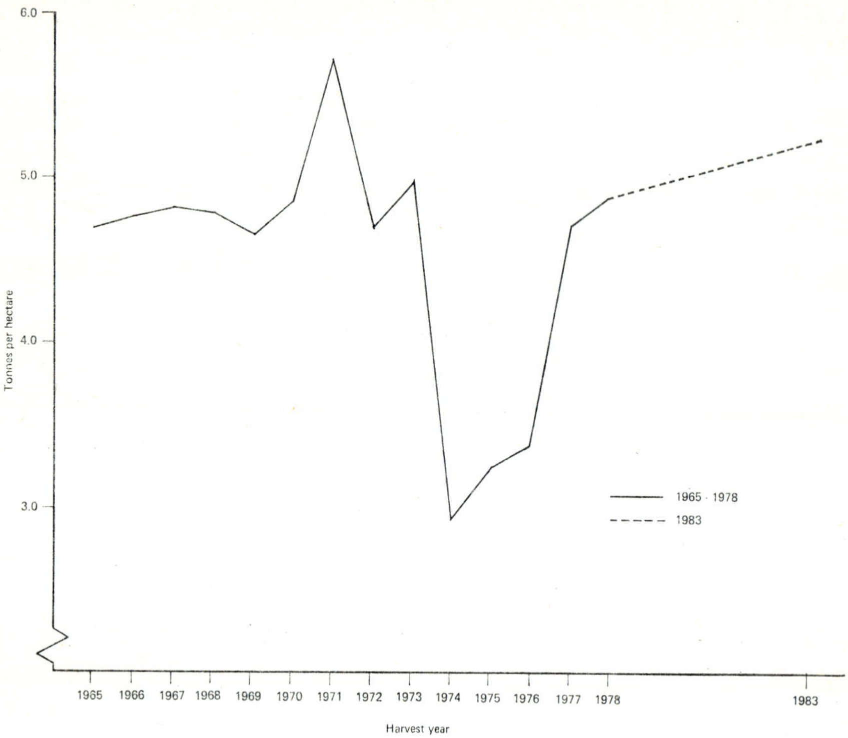

ESTIMATED AND PROJECTED ANNUAL AVERAGE YIELDS OF POTATOES IN THE UNITED KINGDOM

GRAPH 4

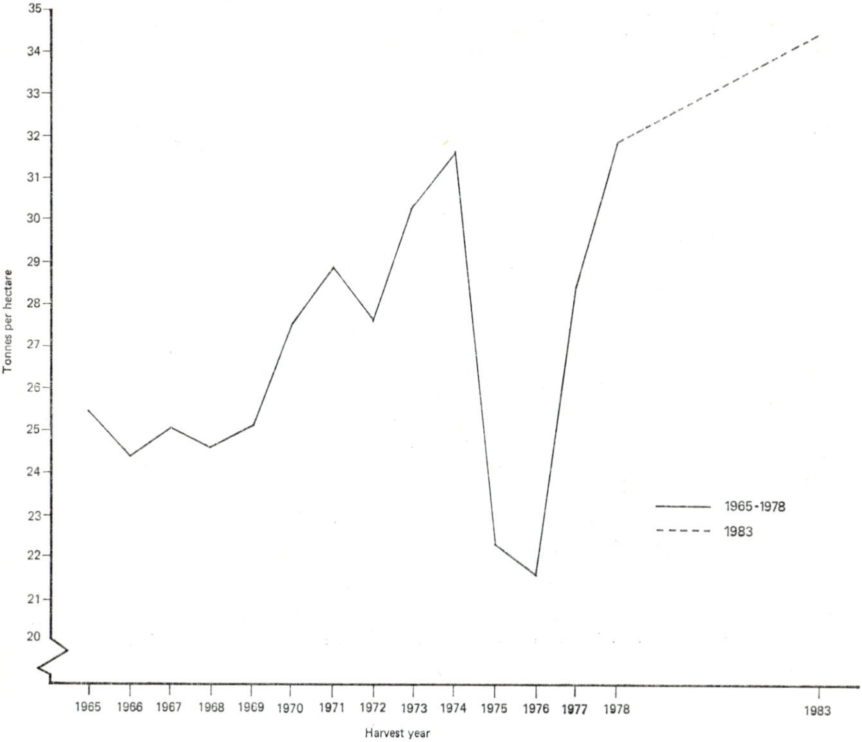

GRAPH 5

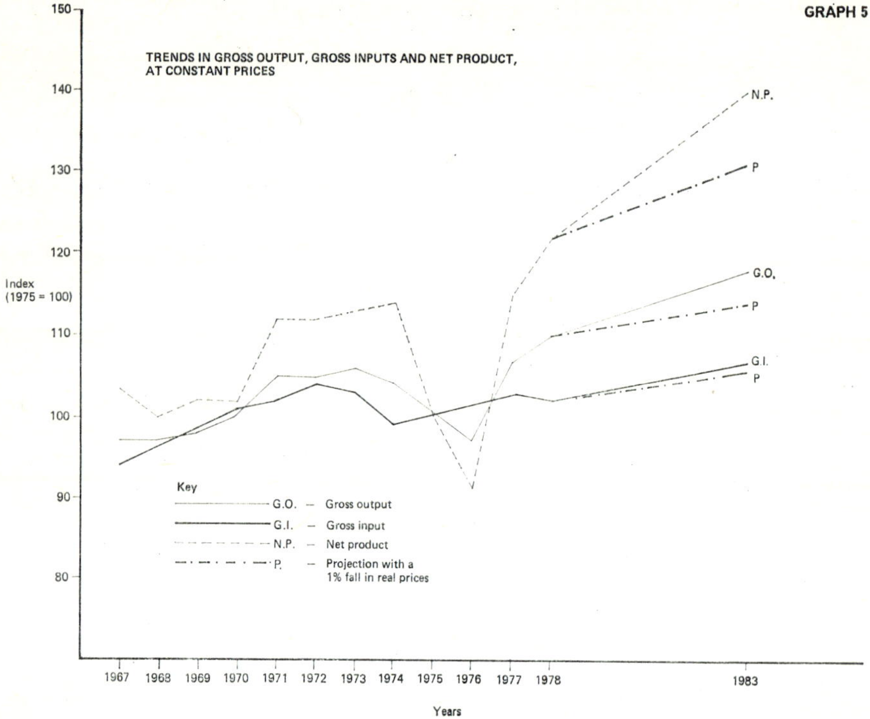

Printed in England for Her Majesty's Stationery Office by Oyez Press Limited
Dd 587021 K48 1/79